NATIONAL GEOGRAPHIC

Ladder

Welcome to India

AROUND THE WORLD

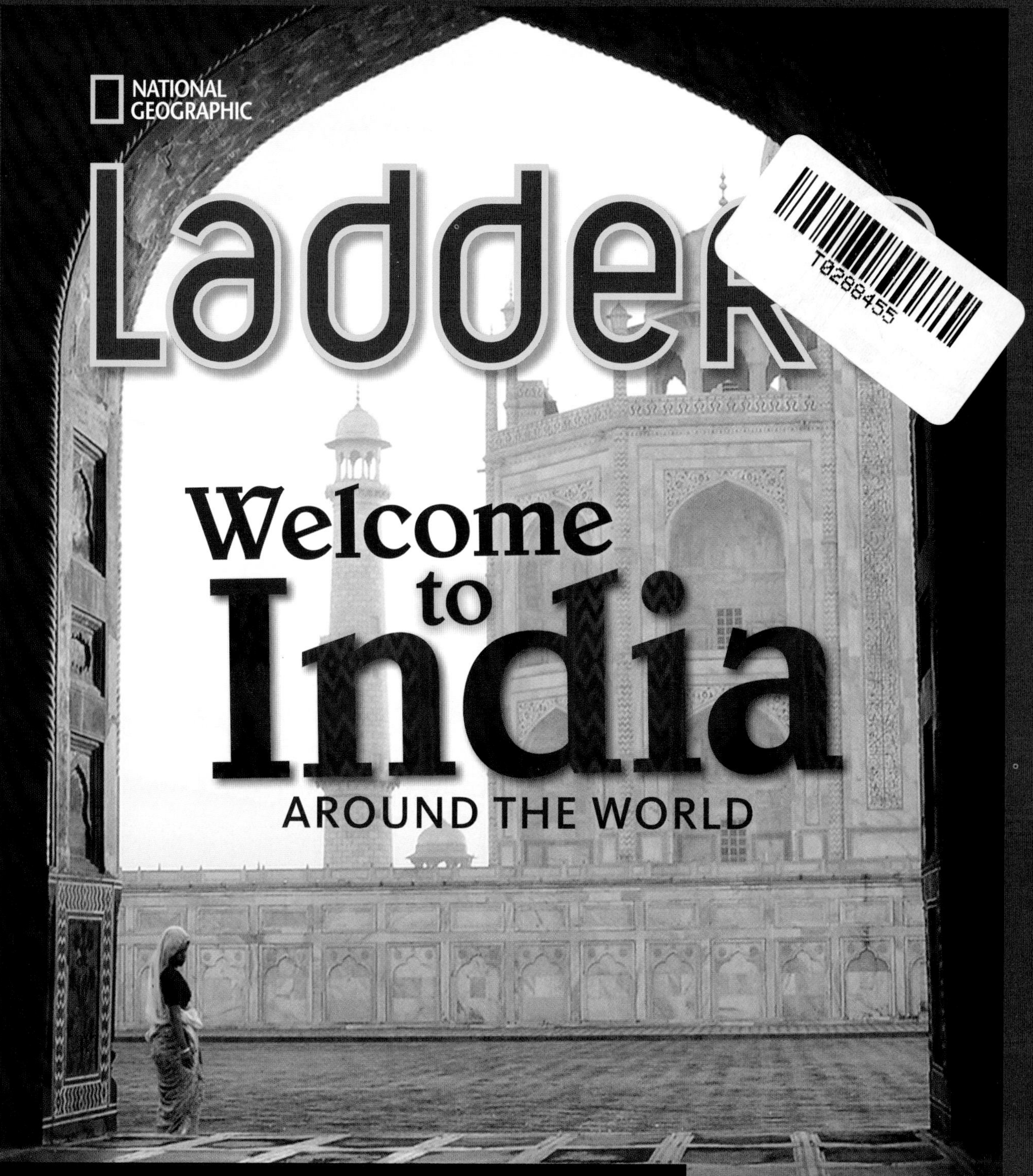

GENRE Social Studies Article

Read to find out about living in villages and cities in India.

WELCOME TO INDIA

by Stephanie Herbek

Namaste! On the streets of India, that's how people say "hello." India is a large country on the continent of Asia. It has mountains, rivers, and deserts—and a lot of people. Over one billion people live in India.

Where a person lives is an important part of his or her **culture**. Culture includes the food, beliefs, music, art, and language of a group of people. In India, the culture can be different in small communities and big communities.

In India, people ride animals alongside cars and trucks. Elephants are a good thing to ride. They have a lot of trunk space!

Small Villages and Big Cities

Most people in India live in small communities called **villages**. Indian villages have simple homes and wide-open spaces to work, grow crops, and play. Plenty of people live in India's large cities, too. These cities are busy, crowded communities. Life in an Indian village can be different from life in an Indian city.

Sometimes people from villages travel to the cities to buy or sell things. People from cities often go into villages to visit friends and enjoy the fresh air. Let's find out more about the villages and cities of amazing India!

Living in a Village

Indian villages are usually made up of groups of small homes in the country built of mud, straw, and stone. Most village homes don't have electricity, running water, or TV, but they keep their owners safe and dry. Besides homes, Indian villages usually have a **well** where people fill their water buckets. There also might be a few stores and a post office in the village. Sometimes there is even a café where people sip cups of tea with their neighbors.

Many villagers are farmers. They grow crops such as rice and wheat to eat and sell. Lucky village families might also own a buffalo that gives them milk to drink. Sometimes these buffalo live inside the house—that's a pretty big pet!

This fishing village is near the water. Check out the colorful boats and homes.

In a village, people start their busy days early. They fill their water buckets at the well, gather wood for the fire, and cook the morning meal. Then most children head off to small village schools. Some children don't go to school because they are busy working on their family's farm. But when school and work are done, all village children come together to share stories and play.

In this village, children play near homes made of wood and grass.

Living in a City

Besides villages, India also has some of the most crowded cities in the world. Have you ever heard of a city named Mumbai (muhm-BY)? Over ten million people live in this loud, exciting city. Mumbai has highways filled with traffic and tall, modern buildings. Most Indians who live in cities live in small apartments in large buildings and work in offices and factories. They do not have space to grow their own food, so they buy it in markets.

In busy Indian cities, people buy their food at outdoor markets.

Most families in Mumbai live in apartment buildings.

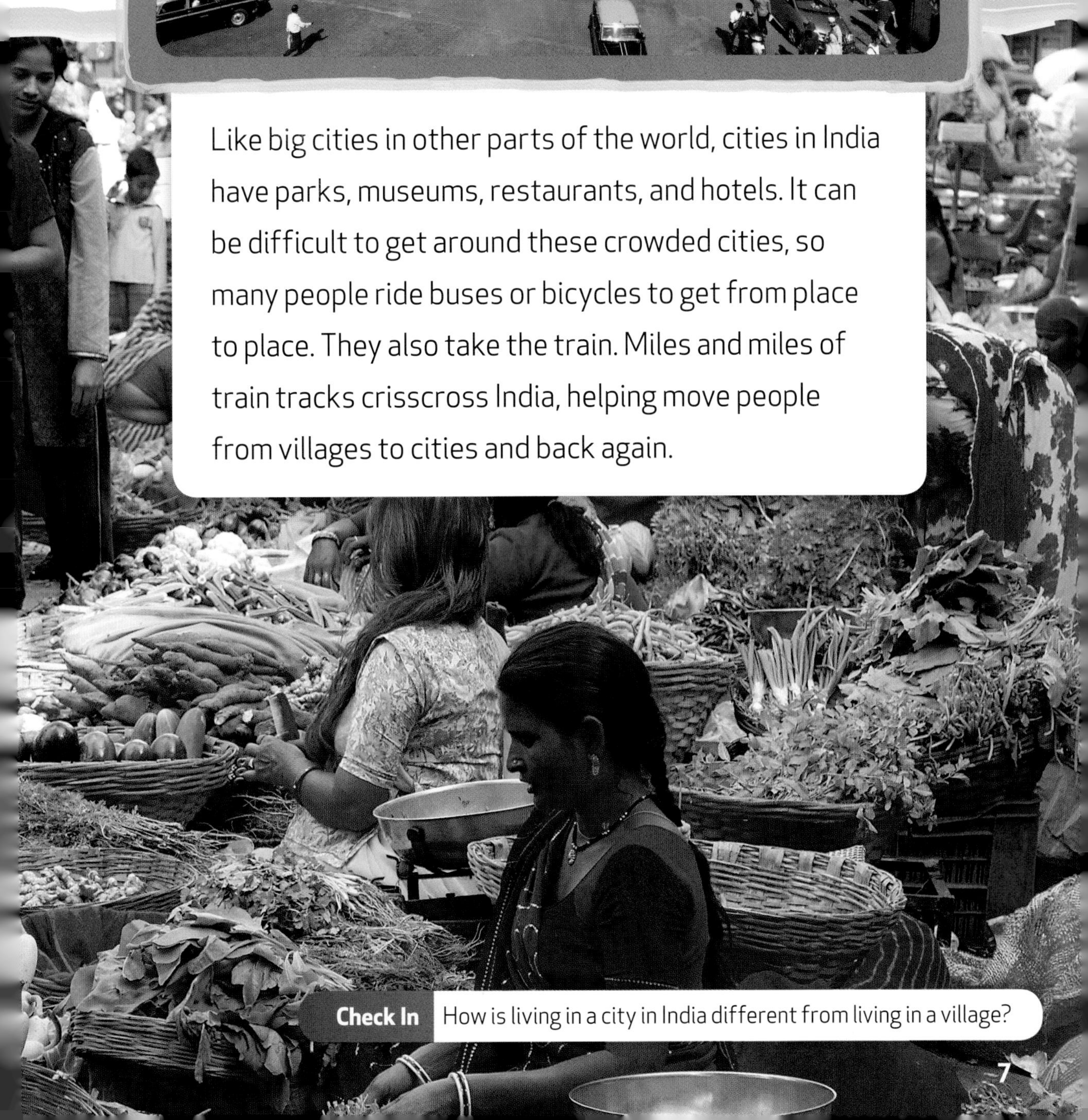

Like big cities in other parts of the world, cities in India have parks, museums, restaurants, and hotels. It can be difficult to get around these crowded cities, so many people ride buses or bicycles to get from place to place. They also take the train. Miles and miles of train tracks crisscross India, helping move people from villages to cities and back again.

Check In How is living in a city in India different from living in a village?

 Read to find out about going to school in India.

Going to School in India

by Grant Williams

For children in India, how you get to school depends on where you live. In cities, getting to school can be as simple as hopping on the bus. But in villages, getting to school can be pretty tricky. Mountains, rivers, and forests stand between some children and their schools. These children may need to ride boats across rivers and tiptoe along swaying bridges just to get to school.

> This boy gets a ride to school each day on his father's bike.

Some Indian schools might surprise you a bit. Sometimes teachers teach and children learn in unusual places. In small villages, teachers and students often hold class outdoors instead of inside a school. In parts of India that are in the desert, school might be held in a tent, away from blowing sand. In crowded cities such as Mumbai, some students can attend classes inside a parked bus.

Many children walk to school in India. Others think it's more fun to run!

Coming through! These children get pulled to school in a little cart.

It's the rainy season! These girls have to wade through flooded streets to get to school.

Learning and Playing

Once they get to school, children are ready to learn. They study many different subjects, including Hindi and English. Hindi and English are the most important languages in India. Many people in India speak Hindi, but learning English makes it easy for Indians to talk to people around the world. It also helps them understand what life is like in other countries. Besides Hindi and English, Indian children also study math, science, and art.

In some Indian classrooms, children sit on the floor instead of at desks.

Like children in the United States, children in India sometimes exercise their bodies as well as their brains during the school day. Many children in India like to play **cricket** and soccer. Cricket is a sport like baseball, played on a large field with a ball and a bat. You probably already know what soccer is. It's a popular sport all over the world!

Small chalkboards, or slates, make it easy to practice writing words in English.

At recess, some children play games with marbles.

Wearing a uniform is usually part of going to school in India.

Lunchtime!

Hungry students can't wait for lunchtime. So, what's for lunch in India? Indian farmers grow crops such as rice, wheat, peas, and beans in the rich soil. Indian children eat many foods made with these important ingredients. Spices are also a big part of Indian cooking, and most Indian people like their foods spicy. **Curry** spices are often used in Indian food, especially to flavor dishes of vegetables or meat served with rice. Chili peppers add some spicy flavor to Indian foods, too.

These children are enjoying their lunch.

Many children in India carry their lunches in a metal lunch box called a *tiffin*. What do they pack? Many different foods! Indian people do not eat very much meat. It can be expensive and hard to find. Rice dishes are easy to bring to school. *Shahi paneer* is a mixture of cheese and vegetables that many children also like to eat. And Indian children bring a bottle of water to school to drink with their lunch. Not all schools have running water.

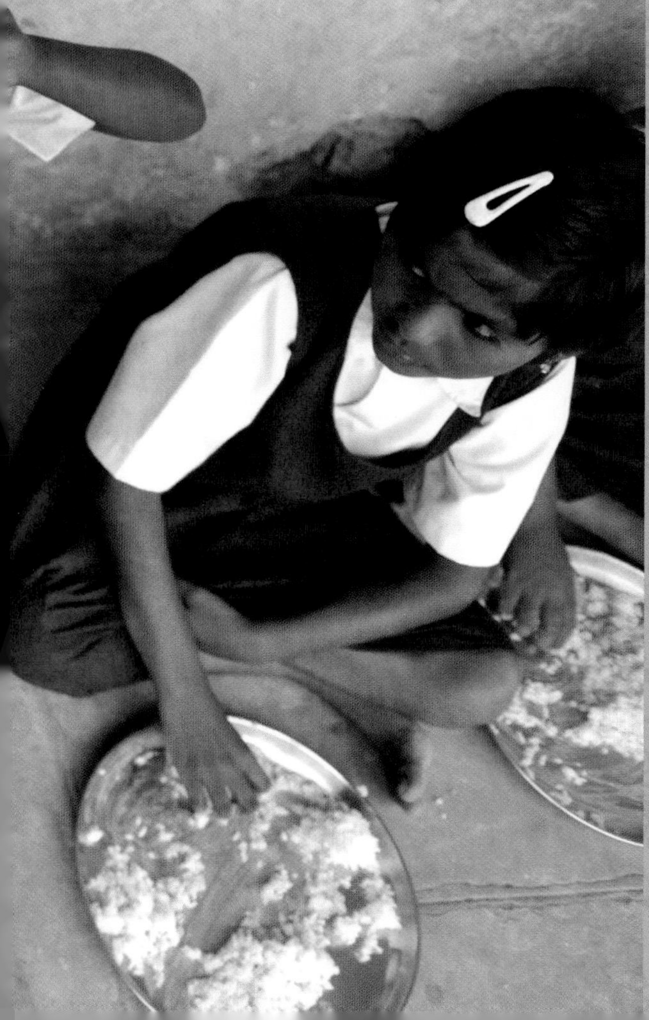

Kids can carry their lunches to school in tiffins.

Naan bread is round, flat, and very tasty.

This delicious curry is made with spicy vegetables and rice.

No School Today!

In India, as in other countries, children sometimes miss school to celebrate important holidays. Instead of going to school, children might spend these days going to parties, eating special foods, and spending time with their family and friends.

Diwali (dih-WAH-lee) is one of the most important **festivals** in India. It is celebrated in October or November. During Diwali, people light up their homes with candles and lamps. Colorful fireworks brighten the night sky, so Diwali is known as "the festival of lights." It is a happy time of year.

Children celebrate Holi by throwing colorful powders at their friends.

Another important festival in India is called Holi (HOH-lee). Holi means "the festival of colors," but can you guess why? Holi takes place every year in March or April and lasts for five days. During Holi, people celebrate the arrival of spring by praying and eating special foods. They also get very playful! An exciting part of Holi is throwing handfuls of colorful powders at your friends and family members. What a messy—but fun—way to celebrate!

Check In What are some of the different ways students get to school in India?

GENRE Folk Tale

Read to find out how rumors get started.

The Foolish, Timid Rabbit

A TRADITIONAL INDIAN FOLK TALE

retold by Jenny Loomis
illustrated by David Mottram

Many stories for children include lessons about life and culture. Jataka tales are one way that children in India learn **morals**, or beliefs about right and wrong. These ancient stories come from Buddhism, a religion that began in India 2, 500 years ago. The moral in this tale explores the danger of spreading **rumors**. As you read this folk tale, think about the lessons that it teaches. Do these lessons sound familiar to you?

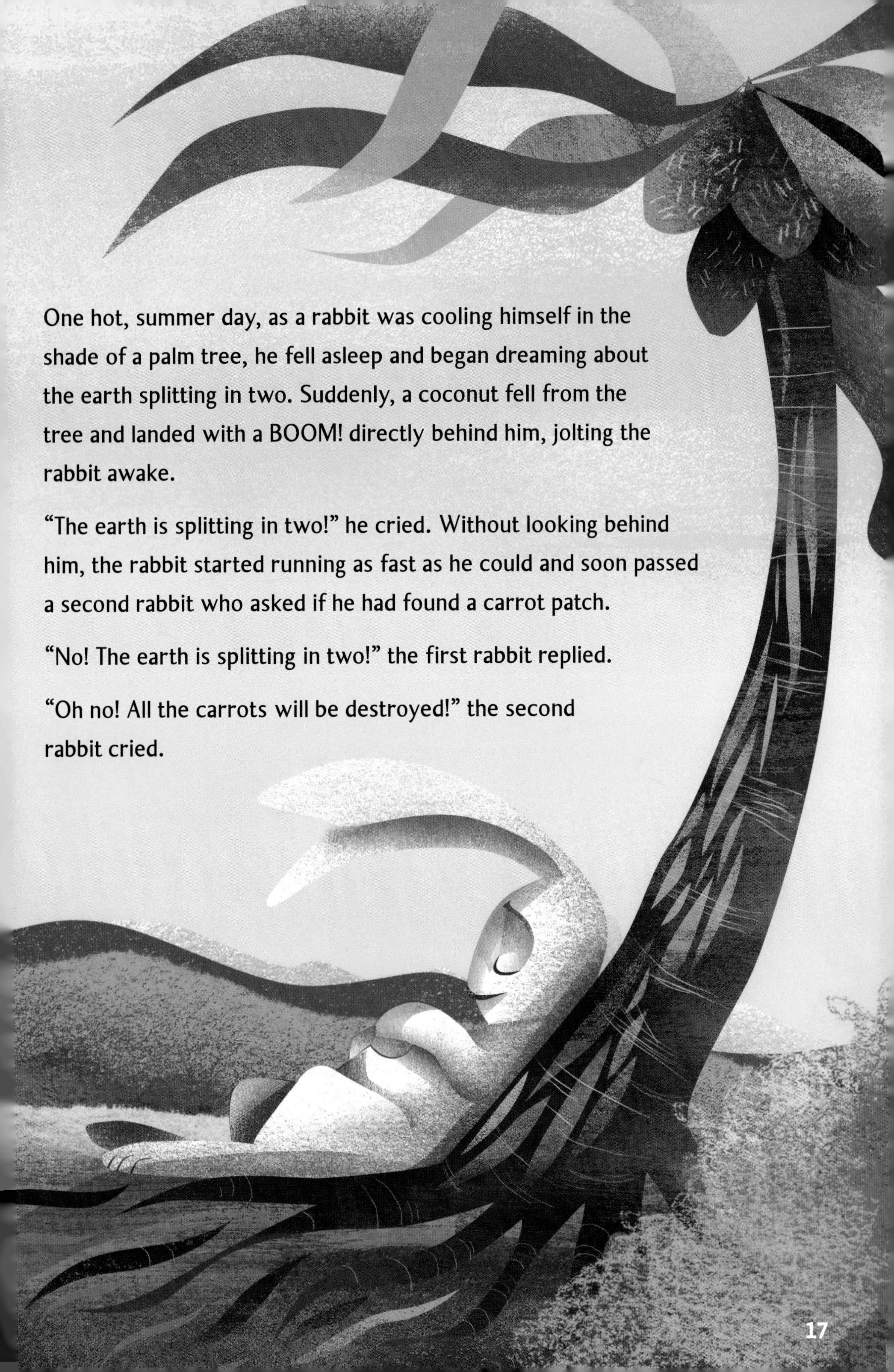

One hot, summer day, as a rabbit was cooling himself in the shade of a palm tree, he fell asleep and began dreaming about the earth splitting in two. Suddenly, a coconut fell from the tree and landed with a BOOM! directly behind him, jolting the rabbit awake.

"The earth is splitting in two!" he cried. Without looking behind him, the rabbit started running as fast as he could and soon passed a second rabbit who asked if he had found a carrot patch.

"No! The earth is splitting in two!" the first rabbit replied.

"Oh no! All the carrots will be destroyed!" the second rabbit cried.

The second rabbit ran as fast as she could to catch up with the first rabbit. The two of them raced through a field, passing rabbit after rabbit, explaining to each that the earth was splitting in two. Soon, there were hundreds of running rabbits.

They ran past a monkey who was sitting in a tree, eating a banana. “Rabbits! Rabbits! Where are you going?” he asked. They explained to the monkey that the earth was splitting in two.

“What will happen to my poor bananas?” cried the monkey, as he swung down from the tree to join them.

The rabbits and the monkey hurtled past an elephant who inquired about the commotion. When the monkey explained that the earth was splitting in two, the elephant said, “How awful! This can’t be good for the vegetation!” He joined the rush of animals, taking care not to step on the rabbits. Then the animals entered a wooded area where they crashed past a deer chomping on some moss.

“What on earth is going on?” asked the deer.

“What on earth, indeed!” answered the elephant. “The earth is splitting in two!”

“Oh my! I hope the moss won’t get hurt!” The deer sprinted to the front of the line of animals.

Nearby, the lion, King of Beasts, was observing the land from the top of a hill. He saw the long line of running animals, was instantly alarmed, and ran to the foot of the hill where he let out a mighty roar. The animals immediately stopped, knowing better than to disobey him. The lion demanded that the animals explain themselves.

"The earth is splitting in two!" the animals yelled together.

"How do you know this?" asked the lion.

"I think the elephant knows," said the deer.

"You may want to check with the monkey," said the elephant.

"Ask the rabbits! They are the ones who told me!" said the monkey, while the rabbits looked at one another in confusion.

"Who started all of this?" bellowed the lion.

The first rabbit timidly raised his paw and admitted that he was the one who discovered the terrible truth about the earth. He described the sound he heard while napping under the palm tree.

"Show me where this happened. The rest of you wait here," demanded the lion.

The rabbit ran back to the palm tree with the lion following close behind.

"This is it, King Lion! This is where it happened."

The lion approached the area near the palm tree and examined it carefully. It was not long before he noticed the coconut.

"Foolish little rabbit, the earth isn't splitting in two. The BOOM! you heard was the sound of this coconut falling from its tree. Let's go back to the others and explain what happened before another stampede occurs."

Upon returning to the hill, the lion explained to the waiting, nervous animals that the earth was quite safe. The animals started cheering, relieved that the earth would remain in one piece.

“But from now on, before acting on a rumor, check to make sure it’s true,” ordered the lion.

All the animals nodded their heads in agreement. This was a sensible plan, they thought. They thanked the lion for his wisdom and assistance. If it wasn’t for him, they might still be running.

Check In What lesson does this folk tale teach?

Discuss

1. Tell about some of the ways you think the three selections in this book are linked.

2. Do most people in India live in small villages or in big cities? Which place would you rather live? Why?

3. Which Indian festival is similar to a holiday or festival that you celebrate? How is it similar?

4. Do you think the first rabbit in the folk tale acted foolishly? What makes you think that?

5. What do you still wonder about the land and people of India?